DATE DUE			

591.56
CAR

Carter, Kyle.

1823

Animals that build homes

RL:
3.7

Quiz # 13050

WALNUT CANYON ELEM SCHOOL
MOORPARK, CA.

427410 01345   45974B   003

ANIMALS THAT BUILD HOMES

THINGS ANIMALS DO

Kyle Carter

The Rourke Book Co., Inc.
Vero Beach, Florida 32964

Edited by Sandra A. Robinson and Pamela J.P. Schroeder

PHOTO CREDITS
All photos © Kyle Carter except page 10 © Tom and Pat Leeson
and page 15 © James P. Rowan

Library of Congress Cataloging-in-Publication Data

Carter, Kyle, 1949-
    Animals that build homes / by Kyle Carter.
    p. cm. — (Things animals do)
    Includes index.
    ISBN 1-55916-112-4
    1. Animals—Habitations—Juvenile literature. [1. Animals—
Habitations.] I. Title. II. Series: Carter, Kyle, 1949-   Things
animals do.
QL756.C386   1995
591.56'4—dc20
                                    94-46845
                                         CIP
                                         AC

Printed in the USA

TABLE OF CONTENTS

HOME BUILDING

Most animals need a home at some time during their lives. Home may be a shelter, a place to raise young or both.

Many animals use homes that they find, such as a cave, burrow or hole in a tree. Other animals are skilled home builders.

Animals of all kinds—from corals and caterpillars to badgers and beavers—make homes. Animals build homes wherever they live—on the ground, underground, in trees, on the water and underwater.

An osprey brings another branch to its growing nest

BURROW HOMES

Burrowing animals live in tunnels that they dig in dirt, sand or wood. Some kinds of termites chew tiny tunnels into wood. Other insects and worms make tunnels in the ground.

Prairie dogs, moles, gophers, marmots, gopher tortoises and burrowing owls are skilled diggers.

The champion earth movers, however, are badgers. These dynamite diggers can shovel dirt faster with their paws than a person can shovel with a spade.

*The badger is the master
of tunnel homes*

HOMES OF MUD

Mud is mushy, so it can be shaped into walls and roofs. When mud dries, it hardens. Some groups of people live in sun-baked mud homes in the desert. Certain animals make mud homes, too.

Flamingoes build cup-shaped nests of mud. The birds use their bills to shape their nests.

The champion "mud wrestlers" are termites. Some of their great, mud mounds rise 20 feet!

This termite mound in Australia
stands about 12 feet high

HOMES OF WOOD

Branches and twigs are handy for animal homes. Bald eagles and ospreys build huge nests of sticks. These birds return to the same nests year after year and keep adding branches and twigs.

Beavers mix branches with mud to build dome-shaped homes. With their long front teeth, beavers gnaw trees down and trim the branches for their **lodges.**

A beaver drags a branch onto its lodge—a dome of mud and sticks

Certain hornets make papery homes

An apartment house in the marsh—Canada geese build a down nest atop a muskrat's lodge

HOMES OF SOFT PLANTS

Animals make homes of grass, leaves or other soft plant material. These nests are well **camouflaged.** They blend in with their surroundings.

The **bittern,** a long-legged bird, weaves a hidden nest of cattail leaves. Another marsh bird, the **grebe,** builds a floating nest of marsh plants. If the water rises, the nest rises, too.

Tent-making bats fold wide leaves into daytime tent homes by chewing along the leaves' spines.

Honduran white bats look down from the roof of their home—a leaf tent

HOMES THAT ANIMALS WEAR

Naturally, animals don't travel in motor homes. However, a few animals do travel in *their* homes.

Snails and turtles, for example, grow a shell. The shell is a traveling shelter, or home.

The box turtle has an amazing hinged shell. The underside snaps tightly shut like a trapdoor. A frightened box turtle pulls in its feet, head and even its tail.

An eastern box turtle tries to decide whether to hide and snap its hinged shell closed

COCOON HOMES

Certain kinds of moth caterpillars wrap themselves in a **cocoon** of leaves or silk. Silk-weaving caterpillars spin a silk "envelope" around themselves.

Tent caterpillars live together in large, silky, cocoonlike tents they build in trees.

Cocoon homes protect caterpillars while they slowly change into adult moths. As caterpillars become adults, they crawl from their cocoons and fly away.

A cecropia moth caterpillar spins a cocoon of silk thread

HOMES IN TREE HOLLOWS

Tree hollows are nesting and resting places for flying squirrels, raccoons, bluebirds, wood ducks, owls and other animals. Few animals can *make* tree hollows, though.

The animal for this job is the woodpecker. The woodpecker's bill is perfectly shaped to chip wood. Also, the woodpecker's skull is super strong to stand up to the hammering.

Woodpecker holes are almost never empty. When the bird moves out, another animal moves in.

Hammering out a nest hole in an old oak tree is just another day's work for a red-headed woodpecker

OTHER ANIMAL HOMES

Animals build homes as different as the animals themselves. Corals build their homes of rock. These small, soft ocean animals produce a liquid that hardens like concrete. The corals live in holes in the rock they make.

Bees make waxy honeycomb homes. Some hornets build basketball-sized nests of a papery material.

Spittlebugs live in bubble homes made of their own **spittle,** or spit. Many penguins build nests of pebbles.

Glossary

bittern (BIH tern) — a long-legged, long-necked wading bird of the marshes

camouflage (KAM o flahj) — coloring that allows an animal to blend into its surroundings

cocoon (kuh KOON) — the covering that certain insects, especially moths, make around themselves just before becoming adults

grebe (GREEB) — a sharp-billed diving bird of ponds and lakes

lodge (LAHDJ) — the dome-shaped home of a beaver or muskrat

spittle (SPIH tul) — spit; clear bubbles of liquid that look like spit

INDEX